# Portable Generators, Understanding and Repairing Them

# Introduction

The tools used in this book belong to me. Therefore it is legal for me to use the pictures in this book.

The information I am supplying is from actual experience. I have tried to combine enough information in this book so that you will understand how electricity and generators work. Understanding how something works enables you to better diagnose and repair that item.

There is a lot going on in a generator. You have the mechanical end or engine to run the generator. Then you have the generator itself. The engine has to be running in order for you to check the generator end. So it makes sense to start with the engine side.

I have provided you with some terms and their definition. Also there is a list of tools and appliances that show their starting and running watts. This information will help in choosing a generator to run what you need to run when using the generator.

Then we will go over how a generator works and how to diagnose problems. I want to give you as much information as possible. If you have some experience then you can use what information you need.

In most cases I will be using pictures when explaining a procedure or identifying something.

# Chapter 1
## Parts of a Generator

I will start with identifying general parts of the generating unit.

# 1 is the access end of the generator. #2 is the power distribution panel with receptacles and breakers. #3 is the power cord from the stator to the control panel. #4 is the muffler for the engine.

# 1 is the air filter compartment. #2 is the engine. #3 is the fuel shut off valve and fuel line to the carburetor. #4 is the engine kill switch. #5 is the choke lever for the carburetor.

\# 1 is the engine. #2 is the starter. #3 is the battery. #4 is the generator. #5 engine oil check and to add oil.

This is a different generator picture just to show how some are built differently. #1 is the on/off switch. #2 is the oil level kill switch.

These can be located in a number of different places. #3 is the engine oil check and where to add oil. #4 is the generator. #5 is the engine governor that speeds it up or down according to the load put on it. #6 is the back of the control panel.

# Chapter 2
## Volt Ohm Meter ( VOM)

The volt ohm meter will be used to diagnose a number of electrical problems. It will be used to show volts ac, volts dc, amps ac, amps dc, and ohms or resistance (continuity). Ohms is resistance in the flow of electricity and is also known as checking the continuity. Continuity means that you have a closed circuit or a continuous loop with no breaks in the wire or component. On some vom's there is a diode and capacitor check. You can check a diode and capacitor with the ohm setting of the vom's also. This is to see if there is a break in the path the electrical current is following.

I will explain the symbols on the meter selector dial. The V with the wavy line over it is Volts AC. The V with a solid line and dotted line over it is Volts DC. The A with the wavy line over it is Amps AC. The A with the straight and dotted line above it is Amps DC. The horse shoe looking symbol is the Ohm resistance or continuity. The arrowhead looking symbol is the diode check selection. The symbol above it is the capacitor check setting. The black lead will always go in the com (common) socket. The red lead has two sockets you can use. The socket that will be used mostly is the socket with the V, horseshoe, and miliamps. This is for voltage both AC and DC, ohms or continuity, and miliamps. The socket above it is 10A. This is to test amps less than 10. The only use for this is very, very low current DC. The little m is miliamp. The crazy looking u is microfares. This is a reading used with capacitors.

If you have a VOM, get it out and take a look at the selector settings. If you want to check AC electrical current you have to know what current you are looking for. You have to select the highest current number on the scale so your test current will be under that. If you are checking for 120 volts AC set the selector to 200 volts AC. If you don't have the higher number selected it may blow a fuse in the VOM or burn it up. This will be true for DC as well. AC is very dangerous, so be very careful. This current is dangerous and can grab you and won't let go. Once you make your check be sure to unplug the tool so you won't get shocked when working on it.

The DC volt and amp readings are not as dangerous as the AC volts. You should still dial in that higher number on the VOM selector though.

The two probes or leads are black and red. The black probe is plugged into the com (common) socket. There are two sockets for the red probe. The one used the most will be ACV and Ohms. The other is for DC volts and amps.

If you want to check an 18v battery to see if it is charged, make sure the red lead is plugged in to the DCV socket. Put your selector switch on a number higher than 18v. Touch the black lead to the negative post and the red to the positive post. On a digital meter you will see what the voltage is on the readout, but on an analog meter you will have to check the needle on the proper scale to get your reading. I like the digital meter better.

To check for a broken wire in your cord, you will select the ohm setting and have the red lead in the ohm socket. This setting will be used more than the others in diagnosing electrical problems.

Some VOM's have a diode check selector setting. This will show if the diode is good or broken. It will also show what the current flow is. A diode allows current to move in one direction. It is still basically a wire. We are  only concerned about the diode being good or bad in these repairs. The diode setting is telling you the amount of current flow through the diode. The Ohm reading tells you if the wire is broken or not. There are times when you may need to know how much current is flowing through the diode, but not with these repairs. The vom telling you the capacitor is a closed circuit could or could not give you enough information about the capacitor. Some stores that sell the capacitors may have the machine needed to check your capacitor so you would know for sure it was good or bad.

# Chapter 3
# Definitions

Amperage – The strength of electric current ( amps )

Brush – A conducting element, usually carbon, which maintains a sliding electrical contact between a stationary and moving element.

Current – The flow rate of electricity.

Direct Current (DC) – Electric current that flows in one direction. Created by chemical or electromagnetic induction.

Electro Motive Force – The force that causes current to flow in a conductor.

Ground – A connection between electrical circuit and the earth.

Ohm – A symbol of an upside down horse shoe on a volt ohm meter. A unit of electrical resistance. Also called continuity.

Rectifier – A device that converts AC to DC.

Resistance – Opposition to the flow of current.

Rotor / Armature – The rotating piece in a generator that produces a rotating magnetic field.

Stator / Field – The stationary part of a generator that contains a set of electrical conductors wound in coils over and iron core.

Exciter – A small source of direct current that energizes the rotor through an assembly of conducting slip rings and brushes.

Volt – A unit of electromotive force.

Voltage Regulator – A component that automatically maintains proper voltage by controlling the amount of DC current to the rotor.

# Chapter 4
## Appliance / Tool Starting and Running Watts

|  | Starting Watts | Running Watts |
|---|---|---|
| 1 HP air compressor - | 3600 | 1440 |
| Belt Sander - | 2400 | 950 |
| Air Conditioner - 10,000 Btu's | 3000 | 1500 |

| | | | |
|---|---|---|---|
| Air Conditioner 24,000 Btu's | - | 4950 | 3800 |
| Circular Saw 7.5" | - | 5250 | 2100 |
| Clothes drier | - | 9000 | 3600 |
| Coffee maker | - | 0 | 1300 |
| Color TV 13" | - | 0 | 50 |
| Crock pot | - | 0 | 240 |
| Electric drill 3/8" | - | 1200 | 480 |
| Electric drill 1/2" | - | 1250 | 1100 |
| Food processor | - | 0 | 500 |
| Hair drier | - | 0 | 1250 |
| Miter saw 10" | - | 4500 | 1800 |
| Planer joiner 6" | - | 4500 | 1800 |
| Refrigerator freezer | - | 1350 | 550 |
| Table saw 10" | - | 4500 | 1800 |

You can get an idea from these examples as to what you can run on your generator. You have to consider the maximum draw on the generator. In some cases, depending on your generator, you may only be able to run one item at the time.

# Chapter 5
## Factors in Choosing a Generator

1. Use the previous chart to determine your power need. Check the start and run watts against what you want or need to run.

2. Check the run time estimate. Ten hours would be good for a work day or to get you through the night of a power failure.

3. How many 110v and 220v receptacles do you need?

4. Is the panel weather proof or sealed?

5.  How quiet is it?

6. Dose it have good wheels and handles for moving the generator around.

7. You may have a certain place that you will store the generator. Take measurements so you can see if the generator will fit.

# Chapter 6
## Engine Maintenance and Repair

The worst things for a generator is sitting up unused and ethanol. Sometimes it will sit out in the weather which gets water in the fuel tank. Some water may get in the tank just from humidity and condensing.

For the non-running engine

Fuel left in the tank and carburetor will cause a varnish type of residue deposited on metal parts. The carburetor has very small holes and channels that can become plugged with varnish or impurities. So cleaning the fuel tank and carburetor is very important.

Take the tank off and turn it upside down and dump all the contents out. After that I would put some gasoline in it and slosh it around and dump it out. Let the fuel tank dry and then use your air compressor to blow it out. Hopefully this will get all the trash out. Some fittings that the fuel line attaches to will have a filter screen on it that rises into the tank. Remove that fitting and clean it. Check your fuel line to make sure it is pliable and not hard and cracked.

When you use gasoline that has ethanol in it it can melt certain rubber or plastic parts in the fuel tank or carburetor. If you have to use the ethanol fuel use an additive made for ethanol fuel.

Take your carburetor off and take whatever needle valves it may have out. I use a bucket of carburetor/parts cleaner to soak it in overnight. When I take them out I spray them with carburetor cleaner and blow them out with high pressure air. I find a very small wire and run it through holes and openings. I again spray the cleaner in it and blow it out. Be very careful and wear safety glasses or goggles. The cleaner can get blown into your eyes.

Check your air filter and if you are not sure about it get a new one. A clogged air filter will keep an engine from starting. If an air filter starts coming apart it could get sucked into the carburetor.

Change the engine oil (#5) and make sure the level is correct. If the level is low it will keep the engine from starting. There is a switch on the engine that looks at the oil level and when it is low it will shut the engine down on will not let it start. There is a way to check this switch. It will have two wires coming from it and one will have a connecter inline. Unplug this wire, sometimes it is yellow. In the above picture the low level switch is located above #1 and behind the black cover. These switches can be located almost anywhere. When you start the engine, plug this wire back in, if the engine dies the switch is bad. If it keeps running, the switch is good.

Check your spark plug and clean it. If there is any question just replace it. Make sure you get the same exact plug to replace it. If you put the wrong plug in it could punch a hole in your piston on the upstroke.

If the engine sill does not want to start check the compression. To do this you will need a pressure gauge with a special end that seals in the spark plug hole. You need to research your engine data to see what that pressure should be. Pull the starter rope slowly a couple of times to check the pressure. If the pressure does not match that data of your engine it will not start. There can be a couple of reasons for this. One, the valves need adjusting, or the piston and cylinder is scored.

You have to get the engine running before you can check the generator to see if it is making electricity. You can do some diagnostics, but it really needs to run.

Gaskets are important when putting the carburetor back together. It is important for any part of the engine you are putting together. Broken or stiff gaskets can leak. Sometimes things leak out of a bad gasket but sometimes air is sucked in. If air is being sucked in it can change the way the engine runs or it might not start at all. So check your gaskets.

If your engine is running

You are ahead at this point so you just need to do some checks. Check your oil level and air filter. Some carburetors have a bleeder on the bottom or side of the bottom bowl. Check this for water.

To keep your engine in this running condition there are a few things you need to do when shutting it down for storage. The first thing would be to shut the fuel off to the carburetor. This runs the carburetor dry. Then disconnect the fuel line from the carburetor and drain the fuel tank. When you need to use your generator you will have to put fresh fuel in the tank.

# Chapter 7
## How Does a Generator Work?

First, a generator needs a means of being turned for it to work. In our case we have an engine fueled by gasoline or some sort of gas. A generator takes mechanical energy and changes it into electricity. A horizontal shaft comes out of the side of the engine and has a rotor attached to it. It is placed inside a stator and a case which is bolted to the engine and the frame.

The components of a generator are: the stator, rotor, case, rectifier/voltage regulator, power cable, and the power distribution panel.

There are three ways that a generator runs to make electricity. One is for the engine to run at a constant speed of 3600 rpm's @ 60 cycles. If the engine speed fluctuates so does the voltage it is producing. The second uses a computer to adjust the engine speed for the voltage needed to be produced. The third type uses a governor to speed the engine up as the power load increases. You may see anyone of these in your repair or when choosing a generator.

You may see some generators described as an inverter type. All this means is that the AC current is converted to DC current and then inverted back to AC power. You still get the power you are looking for, it is just produced a little differently than other types of generators.

A generator needs DC power to energize its magnetic field in the rotor. This DC current is supplied by a rectifier being fed from the stator. Either a rotating or static type exciter is used to excite the exciter coil in the rotor. There are two types of exciters: brush and brush less. The primary difference between these two exciters is the method used to transfer the DC current to the generator rotor fields. Static excitation can be from a battery or a capacitor. A capacitor holds a charge like a battery. The brush type exciter is mounted so the brushes contact the rotor slip rings to transfer the DC current to the rotor. These slip rings are also called collector rings.

Each collector ring is hardened steel with a copper outer coating and mounted to the exciter shaft. Two collector rings are used on each exciter and are fully insulated from the shaft and each other. The inner ring is usually negative and the outer ring positive. You can see the red and black wires attached to the brush connectors coincide with the inner and outer rings.

In this picture you can see the rotor shaft protruding from the middle of the stator and case. On the end of the shaft you can see the bearing that the shaft rides on while supported by the end cap to the case. The next two rings that are copper colored are the slip or collector rings. Further to to the left and closer to you are some of the winding that make up the stator. You can see some bare wire used in the coils and some with insulation. The insulated wires are carrying AC current to the power distribution panel. You can see the power cord taking this collective power away from the stator and to the power distribution panel.

What a generator does is move an electro magnet ( rotor ) near a wire ( the stator ) to create a steady flow of electrons ( power ). The rotor generates a moving magnetic field around the stator, which induces a voltage difference between the windings of the stator. This produces the AC current output of the generator.

This picture gives you a better view of the stator and all the wires wrapped around it and where they come out to the power cord. You can also see the heavy coils on the rotor.

An analogy used to explain this is a water pump pumping water. The water pump takes water and moves a certain amount at a certain pressure. The magnetic field in the generator creates electricity in the stator and it moves out into the power distribution panel.

In an electrical circuit the number of electrons is called the amperage or current and is measured in amps. The pressure pushing the electrons is the voltage measured in volts.

Electricity flows from the negative to the positive. The source of electricity must have two terminals, a positive and a negative. The source of electricity whether it is a generator or battery will want to flow out of its negative terminal at a certain voltage. The electrons will flow from the negative to the positive through a conductor (wire). When this current flows between these two terminals, you have a circuit. Put a motor or a light in this circuit and the electricity does work for you.

DC current is direct and moves in one direction. It moves from the negative to the positive at a certain voltage. A nine volt battery is DC and will move its electrons from the negative to the positive at nine volts.

AC power has an advantage over DC power when it is being produced. It can be manipulated to a higher or lower current. In our case we need to manipulate it to 120v @ 60 cycles. That is what most of our items run on. We use a bridge rectifier or regulator to further change this AC current.

Now we will look at a generator from when it starts to its running state.

Generators with brushes and rectifiers

When the engine starts, the rotor attached to the motor shaft starts turning also. Inside the rotor is an exciter winding. There is a certain amount of iron used in making this coil. If this generator has run in the past it should have had a small amount of magnetism remaining in the exciter coil. If it did, this magnetic field will start passing through the stator coils and produce AC current. With a little or a lot of current some is routed through the rectifier and 12 volts goes into the exciter coil and makes a much stronger magnet. When this magnetic field is increased, this starts producing more AC current in the stator. If this small amount of magnetism is not present, the generator will not start generating electricity. I will explain in a later chapter on how to excite your generator.

When we check our power outlets on the control panel they should all have 120v. This is where knowing what kind of generator you have comes in handy. If it is suppose to run at 36000 rpms it should be adjusted to that. If it is a computer control, the computer may be bad. If it has a governor, your idle may need to be adjusted up or the carburetor tuned.

If your generator has a capacitor you may have less trouble when it starts to be generating electricity. I have been told that the capacitors go out more frequently than the rectifiers. Personally I would rather change out capacitors than have a rectifier. But sometimes we can't make that choice. Capacitors can be cheap where you could have an extra on hand. Rectifiers are from $65-$100.

Generators with a capacitor

This generator is made the same as the one we discussed above. The difference is the components that supply the current to the exciter coil on the rotor. This will be the capacitor. A capacitor is like a battery that is being charged as the generator runs. So it retains a charge.

When the rotor starts to spin inside of the stator the exciter coil has power supplied to it by the capacitor. If the capacitor is working you will get electricity immediately. So the capacitor supplies the power to the rotor to make it an electro magnet. This magnetic field spinning around through the stator coils creates electricity. The current flows through the power cord to the power distribution panel.

Your small portable generators are all going to basically work the same. The way it is excited will change. It is so important to research and look at the parts schematic of a generator. Use forums and reviews to get more information about the generator you are interested in. You may be able to weed out the units that have known problems.

# Chapter 8
## How to Check the Rotor, Stator and Receptacles

In this chapter you will need your volt ohm meter and understand how to use it. Having the previous chapter on reading the vom I will assume you know how to set it up and read it.

The above picture is used to aid you in testing the cable to the receptacles. #1 is the plug that connects to the cable coming out of the stator. #2 is the power panel. #3 is the ground wire. #4 is a set of wires that make a circuit. #5 is also a set of wires that make a circuit. Check every combination of pairs to see if you get a reading. If you do, your wiring to the receptacles are good. If not, you have a break in one of the wires or a bad connection. This will require a closer look at each wire section to check its continuity and connection.

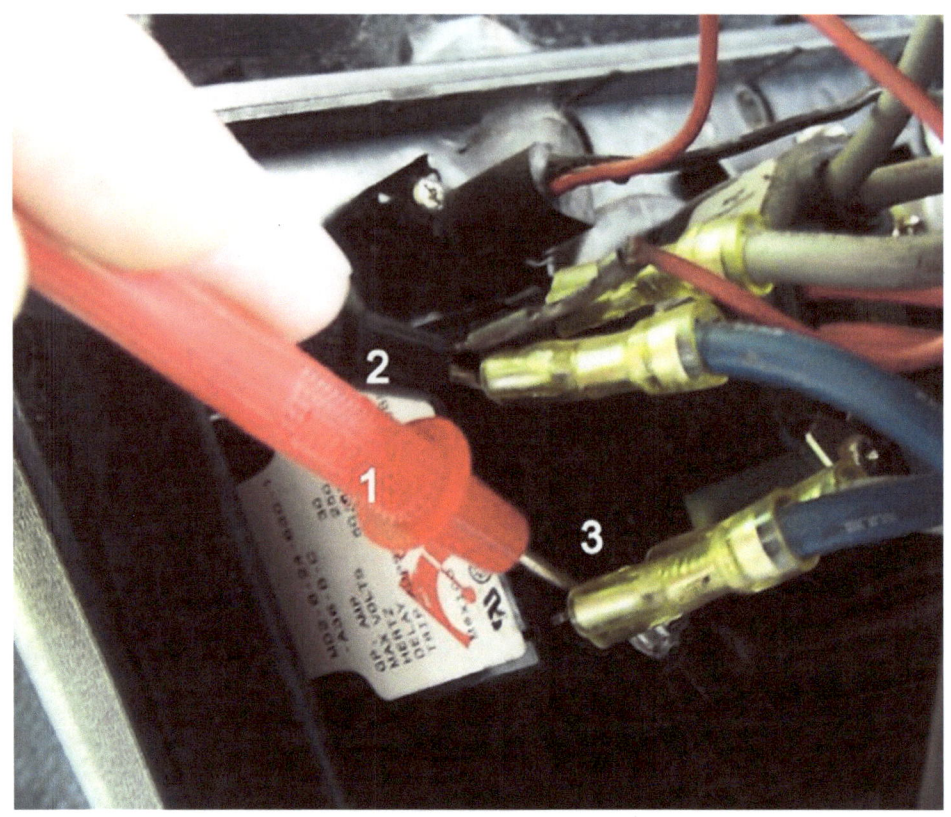

The above picture is the main breaker. Turn the breaker on. #3 is the breaker. I am checking the two connectors to see if I get a reading. If I do, the breaker is good. You should see wires coming in from the power cable and going to the main breaker. These wires are supplying the power coming into the panel. If this breaker is bad, power will not get to the receptacles.

The above picture is of an individual breaker for a receptacle. There will be one of these for each receptacle on the panel. Some of these have push buttons or flipper switches. Make sure it is set and not tripped. When you place probe one to one post and probe 2 to the opposite post you should get a reading meaning the circuit is complete. Check each breaker on the panel. If one shows bad you will need to replace it with the same size amp breaker.

We have now checked the power cord to the receptacles and all the breakers on the panel.

This is the end shot looking into a generator. The center piece is the rotor and the outer wire ring is the stator. You can see the wires coming together to make the power cord at the bottom of the picture. The object behind the hex head bolt is the shaft bearing that sets in a recess in the end cap. This holds the end of the shaft stable and level. The shaft that the rotor sits on is tapered.

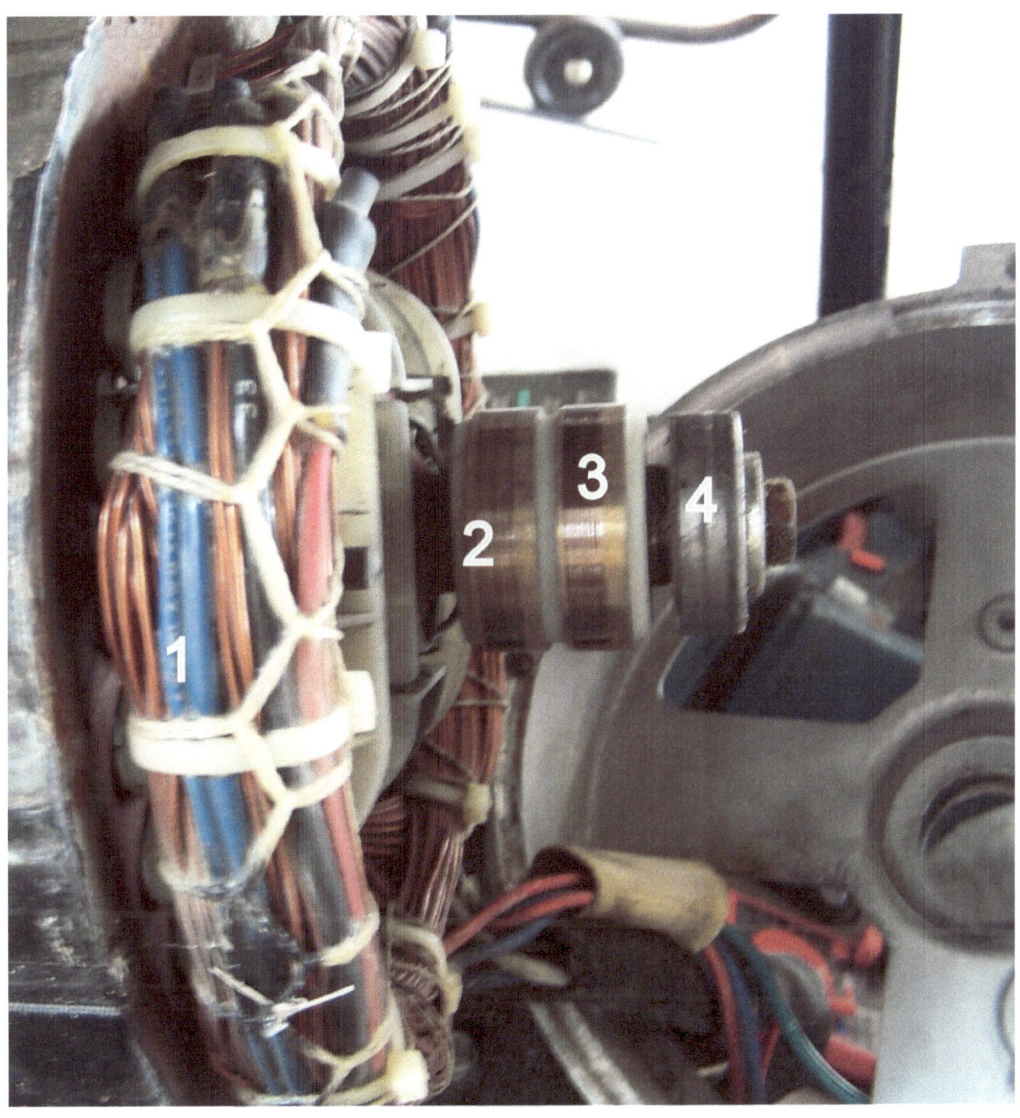

 If you have to take your end plate off this is what you will see. #1 is the stator. #2 and #3 is a collector ring. #4 is the shaft bearing. If you need to check your rotor at this point set your vom to ohms. Put one probe on #2 and the other on #3. If the rotor is good you will see a reading on the vom. If not, you have a broken wire in the rotor. The other check is a ground check. Place one probe from #2 to the generator case. If you get a reading the rotor has gone to ground. Check the #3 ring for a ground. If #2 is grounded, #3 should be also. This will keep the generator from making electricity.

You can also check the rotor by not taking the end plate off. As you can see the two wires connected to the brush holder gives you a place to check with your vom probes. These wires are attached to the brushes which are in contact with the collector rings. Touch your red probe to the red wire and the black probe to the black wire. If you get a reading, the rotor is good. If not, it could have a broken wire in the circuit or there is a problem with the brushes. You will need to remove the brush holder and check each brush. To do this put one probe on the wire connector and the other probe on the brush. Do this for each side. If you get a reading then the circuits are complete. You need to do a ground check also. Touch one vom probe to the red wire and the other probe to the case. If you get a reading this means the rotor is grounded. It will not generate electricity with this ground.

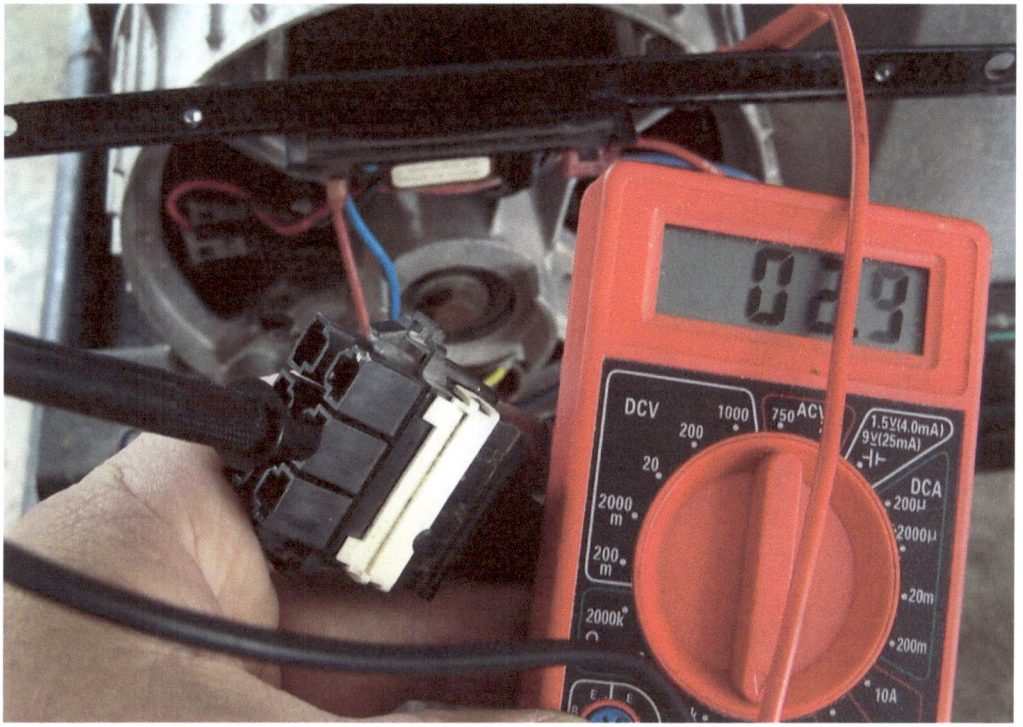

Now we will check the stator

   Locate the ground wire at the plug and touch one probe to the
connector and then one probe to the generator case. If you get a
reading this means your ground circuit is good. If there is no reading,
you have a break in your ground.

In this picture you can see the connectors in the power plug. You can see there is an odd number of connectors with one being the ground. So now we have two sets of wires or two circuits in the stator. Put your probe into one connector and use the other probe to check the other connectors. If you have two good circuits the stator is good. If one of the circuits is bad, the stator is bad. You also need to place one probe on the case and the other probe to each one of the connectors. If you get a reading on one other than the ground, then the stator is grounded. Open circuits or grounded circuits won't produce electricity.

We have now checked the power panel components, the stator and the rotor. If the stator or rotor is bad you will need to check your part numbers so you can check prices. I can tell you that they are expensive and heavy. Freight will also be a factor. If you do the work yourself it may be worth the repair. Every generator is different as is the manufacturer.

Taking the stator off is not a big deal, it is heavy and you may need some help. Remove four bolts and pull it from the engine case and off of the rotor. The rotor is a different story. It is set on a tapered shaft and held in with a long bolt. With the stator out of the way there are two methods to remove the rotor. One is not recommended but if the rotor is bad it won't make a difference. This would be the 2 x4 method. Hold the board at an angle and hit it until the rotor comes off the shaft. If the rotor is good this doesn't seem like a good method unless you are a gambler.

The second method and proper way is a hassle. When you remove the bolt from the shaft the opening is smooth. You need a tap that will fit inside this opening and cut threads. Cut the thread in the end of the shaft. Now you need a piece of all thread, measured from the end of the shaft to the engine to make sure it is long enough with the proper threads. Double nut the end of the all thread so you can turn the all thread into the end of the shaft. The end of the all thread will meet the engine shaft end and push the rotor from the shaft.

Now we have checked three main parts of the generator. If all of these parts check out good, why is it not making electricity. The last item is the rectifier or capacitor. You can check these with the vom but it is really hard to tell if they are good or not. What I suggest is to excite the generator ( see procedure below ).

# Chapter 9
## How to Excite a Generator

We all like to be excited and a generator needs to be excited to make electricity. I will go over how the exciting works automatically. It all starts with a bit of residual magnetism in the rotor. If this magnetism is there it will be a weak electro magnet. As the rotor continues to turn and make more AC current more power flows through the rectifier making the rotor a stronger electro magnet. With the capacitor, unless it is dead, there is a charge that will supply enough voltage to the rotor to make a strong enough electro magnet that it will produce electricity right off.

If either the rectifier or capacitor is bad, no electricity will be generated. We will need to manually excite the generator.

What you need: a 12v battery ( car or motorcycle ), two cables or heavy wires that can carry the load with alligator clips on each end.

The above picture is where the rectifier or capacitor is connected to the rotor through the brushes. As was stated earlier and you can see here the inside ring is negative and the outside ring is positive. Disconnect these wires and attach your cables to the studs on the brush holder. Now connect the other end of the cable to the battery. Negative to negative and positive to positive. Now you have a manual excitation going to the rotor to make it an electro magnet. Put your vom to 200v AC and put the probes into one of the receptacles.

Crank the generator. Now check your vom to see if you have power. If you do, that means the rotor and stator are working. You have manually energized the rotor exciter field and it should have a residual magnetism. Shut down the generator and disconnect the cables from the battery and reattach the wires from the rectifier or capacitor to the brush holder. Now re crank your generator and check your voltage. If you have voltage your exciting worked. If not, your rectifier or capacitor is bad.

Technically you could have two wires sticking out of the generator end cover going to the red and black wires of the brush holder or capacitor. When you crank your generator and touch those wires to a battery the generator should start making power. It would be a little inconvenient, but it would work. If you had parts ordered you could still use the generator. If you were using your generator and suddenly it stopped producing power because of the rectifier or capacitor you could still use it.

# Chapter 10
# Extension Cords

An explanation of how to choose an extension cord follows. The extension cord is sized by the amps of the tool. This is the flow of the current through the wire. If the tool can't get enough current it will burn up. I am listing some average amps per tool and then the cord lengths.

1. Circular saw 12-15 amps
2. Standard power drill 3-7 amps
3. Hedge trimmer 2-3 amp
4. Weed trimmer 2-4 amps
5. Chain saw 7-12 amps
6. Leaf blower 6-12 amps
7. Bug lamp 1-2 amps

8. Lawn mower 6-12 amps
9. Table saw 15-20 amps
10. Sawzall 6-8 amps
11. Router 4-6 amps

Remember that as the wire gets thicker the more electricity, or amperage it can handle. The smaller the number on the cord means the wire bigger. So a #10 gauge wire is bigger than a #16 gauge wire. Just because a cord is thick doesn't mean the wire inside it is. Most cords are written on with the gauge wire or imprinted into the cord. Here is a list of cord lengths and amperage it can handle.

1. 16 gauge cord from 0-100 ft long will handle up to 10 amps.
2. 14 gauge cord from 0-50 ft long will handle up to 15 amps.
3. 12 gauge cord from 50-100 ft long will handle up to 15 amps.
4. 10 gauge cord from 50-100 ft long will handle up to 20 amps.

Some compressor manufactures recommend an 8 gauge wire for over fifty feet. This may be because of a peak pull of electricity or amperage on start up. If you are in doubt, go to the larger gauge to be safe. You can see by the chart, if you went with the 8 gauge you should be safe under any condition. This can be expensive, but you won't destroy your tool.

If you want to run multiple tools on one cord you will have to add all the maximum amps together. If you do that, you can see there are few tools that can be run together at the same time. A homeowner can do this because he will only be able to run one tool at a time. In the construction world where there are multiple people, this will not work. Extension cords are expensive now, but so are the tools that can be burnt up using the wrong extension cord.

# Chapter 11
# Safety

No matter what you are working on you need to be safe. Use safety glasses or goggles to protect your eyes. Use gloves if necessary. If something is too heavy, get help.

Fueling can be dangerous. When gasoline is being poured into a container, it can generate static electricity. The can spout should be resting on the mouth of the fuel tank to ground it. When you are filling your portable fuel can you should take it out of the car or truck and place it on the ground. When you put the pumps nozzle in the mouth of the can, touch the lip of the can. This will ground it and help dissipate the static charge. Gasoline is a dangerous fuel!

Be sure and select a good dry area to set up your generator. You don't want water coming in contact with your control or power panel. Tie your power cords to the frame so they won't be easily pulled out of the receptacle. Use the correct size extension cord for the distance and load you will be supplying power on. The extension cord should not have any cuts or bare spots on it. Even though this is only a portable generator, it is still putting out enough volts to shock or kill you.

Another important part of anything electrical is grounding. We ground electrical panels, receptacles, tools, and generators. A ground is the most effective way to return electricity safely to ground without danger to someone in the event of a short in the circuit. A neutral wire is a return path for unused current. When the short occurs it causes the current to flow through the ground wire causing a fuse to blow or a circuit breaker to trip in the circuit.

Don't take chances!

Extra pictures

This picture shows how the power panel is connected to the power cord coming from the stator.

A different shot of the brush holder and brushes. #1 and #2 are the brushes and you can push them in from the right to the left into the holder. If the spring inside the holder is broken it will not make contact with the collector ring. The complete holder would have to be replaced. You can also check the circuit from #5 to #1 and #4 to #2 with your vom. Put one probe on the left stud and the other on the brush across from it and if you get a reading the circuit is complete. Do the same for the other circuit.

Another shot of the stator wiring and the rotor. You can see in the middle of the picture just to the left of the inner collector ring a silver wire going to one of the rings. There should be another one on the opposite side completing the circuit.

# Epilog

I hope you have learned enough about a generator to understand how they work. Understanding how something works goes a long way in diagnosing problems. There are a lot of parts that make up a generator, but with a vom you can test most of these parts. Circuits make up the generating end to make it work.

A generator doesn't have to be intimidating, the more you learn the more comfortable you get with it. With this books information, your manual, and parts schematic, you should be able to fix your generator. Take your time and take notes if you need to. You can save yourself a lot of money doing your own repair.

The best preventive maintenance you can do with a generator is to run the fuel out of the carburetor and drain the fuel tank before storing. If you do this, you will save a lot of money and your generator will run when you fuel and crank it. I have been using the same generator for eight years. Every time I fuel it up it cranks on the first pull.

When selecting an extension cord check the chart in chapter 10, it may save your appliance or tool. Not supplying enough voltage to your tool or appliance can burn it up. If you are already in a crisis, don't make it worse by using the wrong size extension cord.

I hope you are successful in all your repairs and I hope I have helped you understand portable generators.

# The End